For Ascott-under-Wychwood
Pre-school, where my children
spent many happy days
B. L.

For Finley.
Love, Mummy xxx
K. M.

First published 2022 by Nosy Crow Ltd
The Crow's Nest, 14 Baden Place, Crosby Row
London SE1 1YW
www.nosycrow.com

ISBN 978 1 78800 811 2 (HB)
ISBN 978 1 83994 172 6 (PB)

Nosy Crow and associated logos are trademarks and/or registered
trademarks of Nosy Crow Ltd.

A CIP catalogue record for this book is available from the British Library.

Printed in China
Papers used by Nosy Crow are made from wood grown in sustainable forests.

10 9 8 7 6 5 4 3 2 1 (HB)
10 9 8 7 6 5 4 3 2 1 (PB)

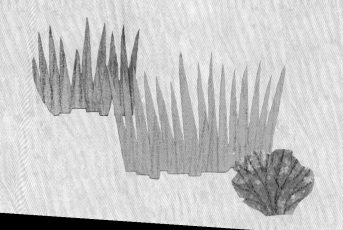

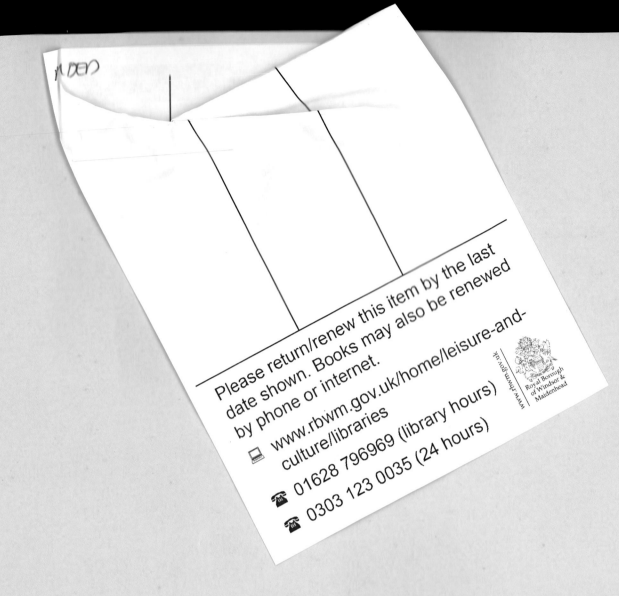

Please return/renew this item by the last date shown. Books may also be renewed by phone or internet.

💻 www.rbwm.gov.uk/home/leisure-and-culture/libraries

☎ 01628 796969 (library hours)

☎ 0303 123 0035 (24 hours)

Royal Borough of Windsor & Maidenhead

www.rbwm.gov.uk

Ben Lerwill • Katharine McEwen

Do Baby Elephants Suck Their Trunks?

Amazing Ways Animals Are Just Like Us

nosy crow

There are babies everywhere!

Some babies can fly.

Some babies
can swim.

And some babies can run faster
than you'll ever be able to –
even when you're fully grown.

Animal babies can be very different to you.

But there are a lot of special ways that you might be the same.

Especially when you were very little . . .

Have you ever sucked your thumb?

Elephant babies, or calves,
sometimes suck their trunks,
just like children sometimes
suck their thumbs.

They also use their trunks to
explore the world, by touching and
discovering new things – just like
you use your fingers and thumbs.

Are you ever carried
around by a grown-up?

Baby orangutans need to be carried everywhere,
just like babies before they can walk. For the first
two years of their lives, they hold tight to their
mothers as they move around the trees.

Some are still being carried
when they're five years old.
The fathers don't help at all!

Where do you like exploring?

Just like human babies as they grow, polar bear babies, called cubs, are very curious about the world.

When they first leave their den, bear cubs learn about
where they live by sniffing, climbing and exploring.
Their mothers teach them how to hunt.

Did you wobble when you started to walk?

Giraffe calves are wobbly when they first try to walk too. They start trying to take steps when they're only a few minutes old, but their long legs make it awkward.

They learn quickly and can usually walk within a few hours – much quicker than human babies!

What were your favourite toys when you were very little?

Baby cats, or kittens, love toys and games.
They use their front paws to play with
balls, soft toys and other small objects.

Like most babies, kittens
have lots of energy,
and playing keeps them
happy and healthy.

How do you like to stay warm?

All babies need to be kept warm when they're born –
especially baby rabbits. They have no hair when they're first
born, so it's important that they stay somewhere cosy.

Mother rabbits usually make their baby nests out of fur and dry grass.

Do you think you drank a
lot of milk as a baby?

Whale calves need a lot of milk too. They're fed underwater by their mothers and drink enough milk every day to fill two bathtubs!

When they're born, humpback whales are about the same size as grown-up hippos. It's no wonder they're thirsty!

How did grown-ups keep you clean when you were a baby?

Baby lion cubs are kept clean by older lions, just like grown-ups look after you.

After eating a meal, cubs often have their heads and faces licked by their mothers or other lions in the pride. A lion's tongue is rough, so it cleans the cub's fur.

Are any of your
teeth wobbly yet?

Puppies are born with baby teeth too.
Young dogs have a set of baby teeth that fall
out as they grow older, just like yours do.

They usually start losing
them at about four months
old, then their grown-up
teeth appear.

Do you think you slept a lot as a baby?

Just like you, koala babies, called joeys, need a lot of sleep. Koalas are very sleepy animals, and the babies sometimes doze off for 20 hours a day!

They live inside the pouches on their mothers'
bellies for six whole months before they are big
enough to explore, clinging to their mothers' backs.

All animals have babies, and in many ways they're just like you were when you were little.

The things that make us the same are as special as the things that make us different.

If you could be any
other kind of animal,
what would you
choose to be?